COLUSA COUNTY FREE LIBRARY
3 0738 00077 0254

D0119125

Everything You Need to Know About *Food Additives*

It is often difficult to determine which food additives in our groceries are unhealthy or unsafe.

Everything You Need to Know About *Food Additives*

Chris Hayhurst

The Rosen Publishing Group, Inc.
New York

Published in 2002 by The Rosen Publishing Group, Inc.
29 East 21st Street, New York, NY 10010

First Edition

Library of Congress Cataloging-in-Publication Data

Hayhurst, Chris.
Everything you need to know about food additives / by Chris Hayhurst. — 1st ed.
p. cm. — (The need to know library)
Includes bibliographical references and index.
ISBN 0-8239-3548-5 (library binding)
1. Food additives—Juvenile literature. [1. Food additives.]
I. Title. II. Series.
TX553.A3 H39 2001
664'.06—dc21

2001001980

Manufactured in the United States of America

Contents

Introduction

The next time you get the chance, pay a visit to your refrigerator. Open the door. Look inside. What do you see?

The answer, of course, is food. But look a little more closely. What's really in there?

Believe it or not, when it comes to food, what you see is not always what you get. That lettuce in the plastic crisper drawer? It may have been sprayed with chemical pesticides to keep bugs from eating it in the field, and unless you've washed it very well, those chemicals are probably still on the leaves. That half-eaten block of cheese? It probably contains rennet, which comes from the stomach lining of baby cows and is used to curdle milk, an essential part of the cheese-making

When it comes to the food you eat today, what you see is not always what you get.

process. And that thinly sliced pound of lunch meat? Analyze a slice and you might find sodium nitrite, a chemical compound used by meat processors to preserve color and taste.

These extra ingredients—things you can't always see and don't always know about—are called food additives. A food additive is any substance or mixture of substances added to food during its production, processing, treatment, storage, or packaging. Additives are additional ingredients. Our various foods contain more than 5,000 additives. Their presence in food may or may not be intentional. Some additives are used to change a food's color or maintain its natural color. Other additives prevent food from spoiling or increase a product's nutritional value. Still others add flavor or smells. Unintentional additives, on the other hand, may include parts of packaging materials that come in contact with the food. Or they may be chemicals used by farmers in the growing of certain ingredients. Some foods have as many unintentional additives as they do intentional additives.

Most experts in the field of nutrition agree: The fewer additives in your diet, the better. In practical terms, this means reducing the amount of processed, or preprepared, foods you eat and instead eating whole foods—foods that are in their natural state with nothing else added. Whole foods are things like fresh fruits and vegetables. Dried beans, rice, oats, nuts, and pasta are also whole

Nutritionists recommend eating whole foods—foods left in their natural state, unprocessed and additive-free.

foods. So are countless other edibles. If your family makes meals from scratch, you're eating whole foods. If you eat out at McDonald's three times a week, or heat up frozen TV dinners, or get your meals premade from a box, then you're consuming a fair share of additives.

There's really nothing wrong with eating additives if you know what they are, what they do, and why they're in your food. But you don't want to eat them unknowingly. It's important to be an educated consumer and not to just accept whatever the world's food manufacturers offer. Be a smart eater. Take control of what goes down the hatch.

Chapter 1

Kinds of Food Additives

A food additive is any substance found in food that isn't a part of the normal ingredients. Additives are used in the production, processing, treatment, packaging, transportation, and storage of food. There are many different kinds of food additives.

Direct additives are added to a food for a specific purpose. Food sweeteners are direct additives. They are added to food to make the food sweeter. Color additives are also direct additives. They change or create the color of a food.

Indirect food additives result from packaging, storage, or handling food. An example of an indirect additive is food-packaging materials or chemicals that accidentally end up in the actual food. For this reason, packaging companies are required to prove to the U.S. Food and

Drug Administration (FDA), the government agency responsible for food safety, that all the materials in their packaging that make contact with food are safe.

Color Additives

Food manufacturers use color additives to improve the color of a food product and make it look more appealing to consumers. Because food processing often leaves the finished product looking bland and stripped of any natural colors, manufacturers must do what they can to get those colors back. Imagine a gray salad dressing or a strawberry shake that doesn't look like it was made from strawberries. Consumers can be finicky. If something doesn't look right, they probably won't buy it.

Two commonly used color additives are dyes and lakes. Dyes dissolve in water and often come in powder or liquid form. They are added to sodas and other drinks, dry mixes, baked goods, confections, dairy products, pet foods, and other moisture-rich products. Lakes are perfect for coloring products containing fats and oils, or foods lacking enough moisture to dissolve dyes. Common lake-containing foods include cake and cookie mixes, hard candies, and chewing gums.

Color additives come in two major categories: artificial and natural. Artificial colors are man-made. They must be certified by the FDA as safe for use in food. Scientists create artificial colors using combinations of laboratory chemicals, then test them on animals for purity and safety. By studying how animals react to the chemicals, scientists can predict how people will react. Following their tests, color manufacturers submit their results to the FDA, which makes the final decision. There are relatively few certified colors in the United States. One example is FD&C Yellow No. 6. It can be found in cereals, baked goods, snack items, and many other processed foods. One advantage of artificial colors is that they can often be made cheaper, purer, and with a more consistent quality than natural colors.

Natural colors—those that come from natural sources like vegetables, minerals, or animals—do not have to be certified. Still, like artificial colors, natural colors must meet certain purity standards before the FDA will allow their use in foods. An example of a natural color is titanium dioxide. This chemical occurs naturally in minerals and is used to give processed candies and frostings, or icings, their white color. It's also used in nonedibles like house paints, eyeliner, and lipstick.

Flavors

If you look at the ingredients of a processed food or drink, somewhere near the end of the list you'll probably see the words "natural flavors" or "artificial flavors." Unlike whole foods, most processed foods would taste

Many breakfast cereals contain artificial colors, such as FD&C Yellow No. 6.

terrible or like nothing at all if it weren't for flavor additives. The food-processing "process," so to speak, is so harsh on the ingredients that any original flavors are quickly destroyed.

Flavor additives are so important to the success of processed foods that an entire industry has been created just to invent and perfect new flavors. Scientists working in flavor factories mix and match chemicals until they get exactly what they want—that perfect taste of french fry, hamburger, or vanilla shake, for example.

Most consumers believe that natural flavors are healthier than artificial ones, but the truth is, natural and artificial flavors are essentially the same thing. They often contain exactly the same chemicals and are manufactured at the same chemical factories. The only difference is that they originate from different sources.

Natural flavors are chemicals that are taken directly from a natural source, like fruits, vegetables, beef, or herbs. For example, scientists use laboratory techniques to extract the chemicals that are responsible for vanilla flavor from Mexican vanilla plants, then transfer that flavor to whatever food or beverage needs it. A similar artificial flavor is vanillin. Used in things like ice cream, pudding, and margarine, it tastes and smells just like real vanilla, but it is made by combining chemicals in the lab.

Recent Ingredients for a McDonald's Grilled Chicken Breast Fillet

Boneless, skinless chicken breast fillets with rib meat containing up to 12 percent of a solution of water, seasoning [salt, sugar, garlic powder, onion powder, spices, whey, flavor (maltodextrin, natural flavors, dextrose, monosodium glutamate), partially hydrogenated soybean and cottonseed oils, Romano cheese (milk, cheese cultures, salt, enzymes, calcium chloride), nonfat dry milk, disodium phosphate], Parmesan cheese powder [enzyme modified Parmesan cheese (milk, cheese cultures, salt, enzymes, calcium chloride), nonfat dry milk, disodium phosphate], xanthan gum, cheese flavor [dehydrated cheddar cheese (cultured pasteurized milk, salt, enzymes), maltodextrin, autolyzed yeast extract], extractives citric acid, soybean oil, and sodium phosphates.

History and Regulations

People have relied on food additives for thousands of years. Salt, for example, has been used to preserve meat since ancient times. Since then, however, much has changed. Many new additives have been invented and many more are in the works. But, unlike the old days, today most countries maintain strict laws and regulations that protect consumers and control what additives can and cannot be used in food. Food manufacturers can't just throw whatever they want into a food and then sell it in the supermarket. They have to play by the rules or not play at all.

The Federal Food, Drug, and Cosmetic Act of 1938

In the United States, the most important law governing food additives is the Federal Food, Drug, and Cosmetic Act. Enacted by Congress in 1938, the act made the U.S. Food and Drug Administration, or FDA, responsible for policing the nation's foods and food ingredients.

The Food Additives Amendment

Twenty years later, in 1958, Congress passed the Food Additives Amendment. This legal change to the Federal Food, Drug, and Cosmetic Act required that manufacturers prove to the FDA that an additive is safe. Only after the FDA gives the additive a thumbs-up, can it be used in foods.

Generally Recognized as Safe Substances

A major exception to the Food Additives Amendment is for substances that are "generally recognized as safe." Generally recognized as safe (GRAS) substances are not required to go through the FDA's regulation process and are exempt from new safety tests. These substances include things like salt, sugar, spices, vitamins, and monosodium glutamate (MSG), as well as hundreds of other additives. They are classified as GRAS substances because they have a long history of being used in food before 1958 or because published scientific evidence shows they are safe, and the FDA doesn't believe it's necessary to conduct new tests. If new evidence ever shows that a GRAS substance may in fact be dangerous, the FDA may review that evidence and take the substance off the GRAS list.

The Delaney Clause

Another important part of the Food Additives Amendment is the Delaney clause, named for the Democratic congressman who wrote it, Representative James Delaney. The law prohibits the FDA from approving any food or color additives that are known carcinogens. Carcinogens are substances that cause cancer in humans or animals. Scientists test substances on animals to see if the substances cause cancer. If a tested animal develops cancer, then that substance is considered carcinogenic.

Color Additives Amendment

Congress passed the Color Additives Amendment to the Federal Food, Drug, and Cosmetic Act in 1960. The amendment is very similar to the Food Additives Amendment. It says that before color additives can be used in foods, they have to first be approved by the FDA.

Good Manufacturing Practices

Good Manufacturing Practices, or GMPs, are government regulations that set a limit on the amount of food and color additives that can be used in foods. GMPs require that when manufacturers use additives, they avoid using more than is necessary to achieve whatever it is the additive is designed to do.

Additive Approval

Food manufacturers can't just throw a new additive into a product and put it on the supermarket shelves. They must first go to the FDA and ask the agency to approve it. The FDA reviews about 100 new potential food and color additives every year.

When a company presents a new additive to the FDA, it's called petitioning. They petition, or present their case, for a new additive. The company must show the FDA that their new additive does exactly what it's supposed to do. To do this, companies perform scientific experiments and collect data. They record their findings and present the results to the FDA.

Scientists must show that new food additives are safe before the FDA approves them for public consumption.

Scientific experiments often involve using animals in situations where using humans might be dangerous. By giving animals large doses of the additives over a long period of time, scientists attempt to prove that those additives would not harm humans. Other studies use human volunteers. The FDA considers the results of these studies when deciding whether or not to approve an additive for future use.

Among other things, the FDA looks at the composition of proposed additives. It considers the substance's properties and how much of the additive is likely to be consumed. It also ponders the long-term effects and other safety issues. All in all, the FDA takes all the scientific data available and tries to determine whether an additive is safe for its intended use.

When the FDA finally approves an additive, it creates rules governing the use of that additive. The FDA decides exactly what kinds of food the additive can be used for, how much can be used, and how it must be identified on food labels. If an additive is intended for use in meat or poultry, the United States Department of Agriculture (USDA) must also approve it.

Once an additive is in use, the FDA doesn't just forget about it. It continues to analyze use of the additive and consumers' consumption of foods containing the additive. It also reads new research or studies that might raise safety concerns. Finally, the FDA's Adverse Reaction Monitoring System (ARMS) keeps tabs on any complaints by consumers or their doctors. If a consumer eats a food containing a specific additive and has an adverse reaction or any other complaints, that information goes into the ARMS computerized database. The FDA uses the database to determine if these reactions represent a true threat to the public's health. If it is decided that they do, the FDA may take an additive off the market.

Chapter 2

Why Food Additives Are Used

Processed foods could not exist without additives. It's a fact of life. Additives keep foods from going bad on the shelf. They ensure that certain foods remain safe to eat despite the threat of dangerous microorganisms. They make food taste the way people expect it to taste. Finally, they allow food manufacturers and distributors to create perfect colors, perfect textures, and perfect flavors. People demand perfection, and additives allow food makers to give us what we want.

According to the FDA, additives are used in foods for five major reasons: To maintain product consistency, to improve or maintain nutritional value, to maintain palatability or wholesomeness, to provide leavening, and to enhance flavor or impart desired colors.

Product Consistency

Product consistency is an important goal of food manufacturers. If a product is consistent, it has the same characteristics day in and day out. All ten granola bars in a box, for example, should look and taste exactly the same.

Manufacturers rely on many different additives to maintain product consistency. One of the most important kinds of additives used for this purpose is called a texturizing agent. Texturizing agents, including emulsifiers and stabilizers, are additives that give texture to foods or change the way those foods feel in the mouth.

Nutritional Value

Without additives, many processed foods would offer nothing in the way of nutrition. For that reason, manufacturers often will add essential vitamins and minerals to these foods. One popular brand of cereal offers a good example. Under the heading "Vitamins and Minerals" on the back of the box, it lists reduced iron, niacinamide, zinc oxide, vitamin B_6, vitamin A palmitate, riboflavin (vitamin B_2), thiamin mononitrate (vitamin B_1), folic acid, vitamin B_{12}, and vitamin D.

Or take milk. If you're like most teens, you probably drink gallons and gallons of the stuff. You pour it on your cereal in the morning, drink it with a sandwich at

Much of the milk one finds at the supermarket is vitamin D-fortified, meaning extra vitamin D has been added by the manufacturer.

lunch, and slug it down with cookies every chance you get. Well, there's a reason you like milk so much: Your body needs it. Milk is rich in protein, high in vitamins A and D, and chock-full of calcium. It helps you grow and keeps your bones strong and healthy. Few other foods offer so much nourishment per glass.

Most milk you buy at the store is vitamin D-fortified. That means the manufacturer added extra vitamin D to supplement that which was already there naturally. Vitamin D-fortified milk is one of the best sources of vitamin D you can find.

But while vitamin D is certainly good for you, there may be other additives in your milk that are not. The

evidence is somewhat shaky for now, but some scientists believe that drinking milk from cows treated with bovine growth hormone may increase the risk of some forms of cancer. Bovine growth hormone, also known as rBGH, is given to many dairy cows to increase their milk production. Manufactured by a company called Monsanto, it is outlawed in Europe, Canada, Australia, and elsewhere, but it has been used in the United States since its approval by the FDA in 1993.

It is possible to buy milk and other dairy products that are rBGH-free. Just look for organic brands. If milk is certified organic, the cows that produced it were never given artificial hormones.

Palatability and Wholesomeness

When something is palatable, it tastes good. When something is wholesome, it's good for you. Additives that keep a product from tasting bad or losing its nutritional value are called preservatives. There are three types of preservatives: antimicrobials, antioxidants, and antibrowning agents.

Antimicrobials prevent the growth of dangerous microorganisms. Antioxidants prevent foods and flavors from going bad as a result of natural chemical reactions with the air. Antibrowning agents include chemicals like sodium sulfite. They prevent foods from turning brown.

Leavening, Flavor, and Color

Leavening agents help baked goods rise. Without leavening, a loaf of bread would be as heavy as a rock and biscuits would be flat.

Many processed foods lack any flavor of their own. Or if they do have a flavor, it's not the type you would want in your mouth. Flavoring additives fix this problem. Similarly, color additives make foods look more appealing. Without them, most consumers wouldn't give processed foods a second glance.

The Story of Salt

If ever there was a super-additive, the additive of additives, it would have to be salt. In the history of additives, there has never been anything else quite like it. Salt is abundant in seawater and is also found as halite, or rock salt, both beneath and above the ground.

The easiest way to collect salt is by evaporating seawater. All you have to do is collect salty water and let it sit in the sun. Eventually the water evaporates and all that is left is salt. Wash the salt crystals with freshwater, let it dry, and it's ready to go. The other way to collect salt is by mining it from the ground. Salt miners zero in on halite deposits and haul whatever they can collect up to the earth's surface. Then, if the salt is top quality, they grind it up, strain out the impurities, and sell it.

Miners in central Thailand haul loads of salt, the oldest and most widely used food additive in human history.

Salt (also known as sodium chloride, NaCl, or table salt) is used most often as a preservative and for seasoning, and it has been around since ancient times, when people first sprinkled it on fish and meats to keep them from going bad. Today, people shake salt on just about anything that could stand a little added flavor—french fries, rice, chicken, vegetables . . . you name it. It's a very simple substance, consisting of just two essential nutrients: sodium and chloride. The human body needs both nutrients to survive.

But too much of a good thing can mean trouble, and many people eat far more salt than their bodies require. Excessive salt intake can cause the body to retain too much water. This leads to increased amounts of blood in the body and results in high blood pressure. And high blood pressure is a leading cause of hypertension, heart disease, and heart attacks. Too much salt can also cause your body to lose calcium, which is essential for strong and healthy bones.

Most people have no idea how much salt they eat each day. But it's safe to say that if you eat a lot of processed foods, you're probably getting more salt than your body needs. Most processed foods contain large amounts of salt. You might not see the word salt on the label, but you will see salty additives and preservatives—ingredients like monosodium glutamate (MSG), sodium benzoate, sodium caseinate, sodium chloride, sodium citrate, sodium hydroxide, sodium nitrate, sodium nitrite, and sodium phosphate. Watch what you eat, and if you're getting too much salt, cut back on salty foods. Your body will thank you.

Pass the Salt

Iodized salt is salt that has a tiny amount of potassium iodide added to it. Potassium iodide is a source of iodine, an essential nutrient required by the human body.

Most supermarket table salts contain what are called anticaking ingredients. These additives—including silico-aluminate, magnesium carbonate, and sodium ferrocyanide—keep the salt from clumping together. If you want pure salt, read the label. The only ingredient on the list should be "salt." Natural, additive-free sea salt, which some experts believe is the best salt available, may not look perfect, but it's easy to recognize. The grains are pale gray, not bleach white, and are irregularly shaped.

Chapter 3

Controversies

Despite the FDA's assurances that food additives are safe, many people have their doubts. In fact, a number of additives that were once approved by the FDA have since been found to be unsafe for consumption and can no longer be used in food. This is not reassuring to concerned consumers. They wonder what additive-laced foods available today will be outlawed tomorrow. And they fear that certain additives in their diet may someday make them sick.

Of the thousands of additives that are approved for use in food, most are absolutely harmless. Experts disagree about some additives, though, and there are a number of substances the FDA's scientists believe are fine for consumption that other experts say should never be a part of anyone's diet. Substances like butylated hydroxyanisole (BHA) and butylated hydroxytoluene (BHT), for example, both of which are commonly used in cereal, have been linked to cancer by some animal studies. Other additives, say experts, may cause hyperactivity and learning disabilities in children.

There are many unanswered questions about food additives. What is certain is that more research needs to be done. Scientists must continue to study additives and mixtures of additives and their effects on those who consume them.

MSG

MSG, also known as monosodium glutamate, is an extremely controversial additive. Food manufacturers, professional chefs, and home cooks alike use it as a flavor enhancer. MSG itself doesn't taste like much, but when you add it to food, it acts a lot like salt: It shifts your taste buds into overdrive and makes otherwise bland food delicious. In fact, MSG is a kind of salt and even looks like regular table salt.

In 1959, the FDA classified MSG as a GRAS substance based on its safe history of use. Since then, however, the Adverse Reaction Monitoring System in the FDA's Center for Food Safety and Applied Nutrition has received hundreds of complaints about MSG. Newspapers and

Monosodium glutamate, or MSG, once a staple in Chinese restaurants before being strictly regulated, is a flavor-enhancing additive that is not safe for everyone.

television programs have run special reports on people who became ill from dining at restaurants known to use MSG. Countless people claim to have allergies to MSG and refuse to eat any food that contains it.

Over the last thirty years, the FDA has reviewed many scientific studies of MSG. Its goal was to figure out what, if any, health risks MSG really does present. Based on these studies, it concluded that MSG is safe for most people. Major health groups, like the American Medical Association and the World Health Organization, agree. They do recommend keeping MSG intake to a minimum and say that severe asthmatics should avoid it whenever possible.

Still, some scientists and consumer groups believe the risks of MSG outweigh the benefits. They feel the chemical should not be used on foods at all. They point to studies showing that MSG causes brain damage and reproductive disorders in animals. They also feel there is enough real-life evidence to show that people can get extremely sick from eating MSG and that things like chest pain, nausea, bad headaches, and numbness in the neck, arms, back, and face are all too common.

As it is now, the FDA requires that all foods containing MSG be labeled. You'll see the words "monosodium glutamate" in the list of ingredients. Do you want MSG in your food? It's up to you.

Sulfites

Sulfites are chemical preservatives primarily used to keep foods from turning brown. They are added to baked goods, jams, dried fruits, seafood, soup mixes, and many other foods. They're also found in alcoholic beverages. Take a look at the label on a bottle of wine and you'll probably see the words "CONTAINS SULFITES" in capital letters.

Some people are allergic to sulfites. They may develop a skin rash on their chest after drinking wine made from sulfite-treated grapes. They may feel sick after dining at a restaurant salad bar that was sprayed with sulfites to keep the lettuce from wilting. In more serious cases, people with asthma who inhale sulfite fumes can experience severe asthma attacks, which make it very difficult to breathe. If things get severely bad, they can die.

Chances are you'll never have a problem with sulfites. Most people eat sulfite-containing foods all their lives and do just fine. Still, you should be aware of this potentially harmful additive and do what you can to avoid it. Especially if you're asthmatic, be careful. When you eat at a restaurant, ask the manager or your server if any of the food contains sulfites. When you eat processed foods, first read the label. The FDA requires that foods containing sulfites say so on the packaging. Look for ingredients like sulfur dioxide, sodium sulfite, sodium and potassium bisulfite, and sodium or potassium metabisulfite.

America's Sweet Tooth

Of the 160 pounds of sugar the average American consumes every year, almost 60 percent—that's ninety-six pounds—is from corn sweeteners, the stuff that's added to soda to make it taste so good.

There are fifteen calories in one teaspoon of sugar. One tablespoon of jelly and one cup of Kool-Aid each contain up to six teaspoons of added sugar.

Sugar Substitutes

Here's a tough fact to swallow: The average American consumes the equivalent of 160 pounds of sugar every year. Just how, you might ask, is that possible? Well, consider a twelve-ounce can of soda. Crack open just one can, open your mouth, tilt back your head, slug it down, and you've taken in about nine teaspoons of sugar. Taste good? Thought so. Now think back on that bowl of cereal you had for breakfast. Awfully tasty, wasn't it? And that snack before lunch? And that candy bar this afternoon? And dessert this evening? Add it all up. One hundred sixty pounds doesn't seem like such a big number after all, does it?

The truth is, we're obsessed with sugar. Think about it. Cake, ice cream, cookies, soft drinks—all our favorite foods are full of the stuff. There's no doubt about it—sugar tastes good.

Unfortunately, there's a problem. Ask just about any professional nutritionist, and he or she will tell you sugar—especially 160 pounds of sugar—is bad for you. Sugar rots your teeth. It makes you gain weight. Eating excessive amounts of it has been shown to cause heart problems, hyperactivity, weight gain, and dangerous diseases. A little sugar in your cereal may not hurt you, but meal after meal, day after day, year after year, it eventually adds up.

You'll find sugar in many different forms, including glucose, xanthum gum, dextrose, fructose, fructose syrup, fruit juices, lactose, honey, molasses, corn syrup, maltose, rice syrup, and sucrose. Sugars can be used to improve a food's flavor. They can also be used as preservatives. Check labels. If you see any of these ingredients, you'll know the food has sugar added.

Because sugar has such a bad reputation, the food industry has come out with so-called sugar substitutes. Sugar substitutes are artificial sweeteners. Made from chemicals by laboratory scientists, they're far sweeter than regular sugar. Sugar substitutes are usually used in processed low-calorie foods—the kind of foods that dieters love to eat. Because the artificial sugars are so sweet, food manufacturers can use just a fraction of the amount they would need if they used real sugar. That keeps the calorie count down. It also keeps dieters, who are concerned about their weight, happy.

The FDA has approved several sugar substitutes for use in foods, and they may approve more in the future. Four popular substitutes are saccharin, aspartame, acesulfame potassium, and sucralose. Sucralose has enjoyed relatively few problems since it came on the market in 1998, but the others, especially saccharin, have been extremely controversial. Some people feel they should be banned from use in foods.

Here is a quick rundown on the four big artificial sugar substitutes.

Acesulfame Potassium

Acesulfame potassium, also known by its trade name Sunett, was first approved by the FDA for use as a tabletop sweetener in 1988. The chemical is about 200 times sweeter than sugar and has zero calories. According to the FDA, more than ninety studies have been conducted that show that it is safe as an additive in things like baked goods, frozen desserts, candies, beverages, and more than 4,000 products worldwide. Still, some independent groups question the results of these studies and advise consumers to avoid products containing acesulfame potassium at all costs.

Sucralose

You may know sucralose by its trade name, Splenda. About 600 times sweeter than sugar, it was approved by the FDA in 1998 as a tabletop sweetener for use in products like baked goods, nonalcoholic beverages, gum, frozen dairy desserts, fruit juices, and gelatins. Today it's approved for use as a general-purpose sweetener in any processed food. At least 100 scientific studies have been conducted on sucralose, and most people agree it has been proven a relatively safe alternative to sugar.

Aspartame

Aspartame is found in many diet soft drinks, desserts, breakfast cereals, chewing gums, and other products. It's also commonly used as a low-calorie tabletop sweetener in brands like NutraSweet and Equal. Many coffee drinkers pour these packets into their morning brew. The FDA claims that more than 100 laboratory tests prove aspartame is safe as a food additive, but some consumer safety groups don't agree. For one, evidence has suggested aspartame consumption can lead to unwanted side effects like headaches, hallucinations, and depression. Some experts also question the quality of the animal-based cancer tests used in the studies, which they feel were inadequate. For now, aspartame is still approved by the FDA, and there are no plans to ban its use in foods. Still, foods containing aspartame must carry a warning label telling consumers it may harm people afflicted with a rare brain disease called phenylketonuria.

Saccharin

Saccharin is more than 300 times as sweet as sugar. It's used primarily in diet sodas and as a tabletop packet sugar substitute called Sweet'N Low. Most major U.S. health organizations, including the American Medical Association, the American Cancer Society, and the American Dietetic Association, agree with the FDA and give saccharin a thumbs-up when it's used in moderation. Still, saccharin has been the subject of extreme controversy over the last twenty years, mainly because laboratory experiments have repeatedly shown that it causes cancer in rats.

Saccharin was first discovered in 1879 and was used as a food sweetener for almost eighty years before the 1958 Food Additives Amendment became law. At that point, because of its long and safe history, saccharin became a part of the FDA's generally recognized as safe list.

In the 1970s, however, the FDA took a hard look at saccharin and hundreds of other GRAS substances to make doubly sure they were safe to eat. To the surprise

The sugar substitute saccharin, sold in packets under the brand name of Sweet'N Low and found in diet sodas, has a controversial history.

of many scientists and consumers, tests showed that saccharin caused carcinogenic tumors to develop in rodents. In 1977, the FDA took action and decided to ban saccharin from use in food.

The ban didn't last long. Some people questioned the studies and felt that they were misleading. They believed the results of the tests failed to prove saccharin was a true danger to humans. The U.S. Congress responded to the doubts and passed the Saccharin Study and Labeling Act, which overturned the FDA ban while more experiments could be conducted. The act allowed saccharin use to continue but required that all foods containing the substance carry a warning label.

Today saccharin is still a popular additive, although many educated consumers refuse to eat it. And despite tests that continue to show saccharin causes cancer in animals, the substance has not been banned again.

Chapter 4

Alternatives to Food Additives

If you're willing to take the time to be a smart eater, you can easily minimize the amount of food additives in your diet. All you have to do is shop around.

The first thing to do is take an inventory of the food in your house. Open your cabinets and empty the fridge. See what's really in there, and see if you can identify all the additives in each product.

Next, do a little research. Go to your local library and pick up a book with detailed information about individual additives (see the For Further Reading section at the end of this book for suggestions). Make a list of the additives you found in your food at home, and see what you can learn about them.

At this point you're ready to make a decision. Are you OK with the additives currently in your diet? Or is it time for a change?

If you decide you have to take action, have a talk with your parents. Explain to them what you've learned about additives and tell them what you think. Then, next time they go shopping, see if you can go along.

Reading Labels

At the store, as you stroll the aisles, read the labels. By reading labels you'll be able to make smart choices every time you buy packaged food. But don't get ahead of yourself. Labels can be tough to understand. You might even say they're misleading. For example, a product might claim to be sugar free, but it may still have sugar products like xylitol or sorbitol. Read the label carefully, and try to figure out what's really in the package. Know what the additives are and know where each one comes from, how it's used, and any potential health effects. And remember, some additives—like vitamins and minerals your body needs to stay healthy—are good for you. Don't worry if at first it all seems incredibly confusing. Before long you'll have no trouble weeding the additive-full foods out from the additive-frees.

"Natural" Foods

Packaged foods that claim to be natural may or may not be full of chemicals. For the most part, the word "natural" means nothing. But often you'll find that such products are a good place to start if you're trying to find foods low in additives.

Food manufacturers in the so-called natural food industry usually use far fewer additives and preservatives than other food manufacturers. A big reason for this is food quality. Poor-quality foods often need additives to make them look and taste better. High-quality foods—the kinds you may find behind "natural" labels—are often already tasty, full of nutrients, and visually appealing.

Chemically treated and refined table sugar contains almost nothing of value to the human body. Nutritionists say sugar supplies "empty calories." That is, you get fifteen calories for every teaspoon of sugar you consume, but that's all you get. No vitamins. No minerals. Nothing your body really needs.

But not all sugars are created equal. There are natural alternatives to table sugar that are still sweet and tasty, but much friendlier to the body. Next time you reach for the sugar, make it natural. Here are a few options:

- *Maple syrup* **Pure maple syrup is concentrated sap from the maple tree, plain and simple. Put it on your pancakes, add it to your oatmeal, and use it almost anywhere you would use regular table sugar. Your body will thank you for the extra potassium and calcium. Avoid the fake "maple-flavored" syrups, which are unhealthy mixtures of corn syrup and** artificial flavoring additives. Also, get organic maple syrup whenever possible. Otherwise it may include nasty chemicals like formaldehyde and antimold additives.

- *Honey* Bees make honey from flower nectar, and beekeepers collect it from their hives. Honey is very sweet—even sweeter than table sugar. It's not as nutrient-rich as some other natural sugars, but it does contain trace amounts of minerals.
- *Molasses* Blackstrap molasses is your best option. This dark, flavorful syrup is rich in calcium, iron, and potassium. It's most often used in baked goods. Organic molasses comes without harmful pesticides, herbicides, or sulfur dioxide.

Whole Foods

Another great way to avoid additives is by eating whole foods. Whole foods are unprocessed and minimally packaged. They have a much higher nutrient content than processed foods. They contain more fiber, essential vitamins, trace minerals, and healthy phytochemicals like carotenoids, isoflavones, and bioflavonoids, which many experts believe reduce your chances of getting certain types of cancer.

If your family cooks meals from scratch, whole foods are already a part of your diet. And if that's the case, there's a good chance the number of additives you consume are minimal.

CSAs: Guaranteed Additive-Free Food

Food means a lot of different things to a lot of different people. For many, the quality of a meal depends on the time it takes to be eaten. For them, the ideal feast involves food that's easy to eat and saves time—things like frozen TV dinners, microwave pizzas, and fast-food burgers. Unfortunately, most of this food is full of additives and chemicals and is far from "natural." Other people would never dream of rushing a meal. To them, when it's time to eat it's time to think about how the food was grown, who grew it, and the effort that went into preparing it. On their plates you'll find foods such as fresh fruits and vegetables and grains. You'll also find far fewer chemicals.

If you happen to be among this second group of eaters, you'll be happy to know there's a great new way to really get to know your food. Thanks to the growing popularity of food-buying programs known as CSAs, or community-supported agriculture programs, you and your family can buy almost all the food you'll ever need directly from a farm or group of farms near your home.

CSAs create a bond between farmers and hungry consumers. Consumers join a CSA when they give the farmer a set amount of money before the growing season begins. A typical fee might be $350. The farmer uses that money to cover the cost of things like seeds, tractors, and salaries for employees. Later, when the food is finally harvested, CSA members get weekly bags of fresh, wholesome, chemical-free food. All they have to do is go to the farm and pick the food up; or, if they live in the city, they can wait for the food to be delivered to the local farmers' market and pick it up then. They already paid at the beginning of the season, so there's no need to bring money.

Joining a CSA is not exactly risk-free. If the crop fails for some reason, whether because of a lack of rain, plant-eating pests, or any other act of nature, you may lose your money. But if the crop succeeds, like it almost always does, you'll enjoy a variety of fresh vegetables, fruits, herbs, and, in many cases, eggs, milk, meat, and other edibles. Better yet, you might even get to help the farmer out. It's not uncommon for CSAs to ask their members to lend a hand on harvest day, and many CSAs encourage members to pick their own berries or flowers.

One of the best things about CSAs is that most use organic farming methods. Organic farming makes the health of the land a priority. No harmful chemical pesticides, herbicides, or fertilizers are ever used on an organic farm. The goal is to grow fresh food in an environmentally friendly manner.

The truth is, while many people continue to eat without ever questioning what they're eating or where it was grown, many others are turning to CSAs for a healthy and environmentally responsible alternative. By joining CSAs, these people say no to mass-produced and chemically treated foods. They commit to the health of the land and the health of their bodies, and give farmers the message that they're willing to help them make a living in a world where most people are perfectly happy with chemically manufactured processed food. If you think you'd like to make a difference, join a CSA.

Chapter 5

Additives Today

Americans love processed foods. In fact, in this fast-paced world of ours, most people think they could never live without them. We buy processed foods because they are convenient. They are quick and easy to prepare. They make life simple. There's no thought involved. Open and serve is the name of the game.

Thanks to this demand, this need for speed, thousands of new processed-food products swamp supermarket shelves every year. And with them come thousands of additives. There are additives for everything. Nutrition, preservation, flavor, color—you name it, an additive is what makes it happen.

A scientist performs a test in a food and beverage lab, where many types of crops and livestock are augmented with genes from other species.

Genetically Engineered Foods: Are They Safe?

In recent years, many of the foods we eat have gone through major changes. Sure, they may look the same as they did in the past. And for the most part they taste the same. But thanks to a new laboratory technique called genetic engineering, they're very, very different.

In genetic engineering, scientists take certain genes from one organism and give them to another organism. An organism's genes are what determine its traits—things like how it looks and other characteristics. Genetic engineers, as these scientists are known, work to invent new combinations of genes that would never exist in nature. When they combine the genes in the right way, they can create foods with new characteristics. One of the first genetically engineered (GE) fruits available at supermarkets, for example, was the Flavr Savr tomato, which appeared on shelves in 1994. The tomato, which looked normal, included a gene from a cold-water fish that increased its resistance to frost and prevented it from spoiling for long periods of time.

Since the first Flavr Savr, genetically engineered crops have become incredibly popular on U.S. farmland and in the supermarket. Today the majority of food products offered in the United States contain ingredients from GE crops.

The safety of GE foods sold in the United States is determined by the U.S. Department of Agriculture, the Food and Drug Administration, and the Environmental Protection Agency. Still, some people wonder if genetically engineered foods might be dangerous. They worry that the government agencies don't conduct their own experiments to decide if a food is safe and instead rely on test results supplied by the food manufacturers. They wonder if those food manufacturers might "fake" some of their results in order to guarantee that their new genetically engineered foods are successful.

Opponents of GE foods point to a report by the National Academy of Sciences (NAS) that said GE crops have the potential to cause serious health and environmental problems. Potential risks include harm to insects and organisms in the soil, the creation of "superweeds" immune to natural and chemical herbicides, and increased resistance to pesticides by insects. The report also said that genetically engineered food may increase the risk of allergic reactions in some people. The NAS report did point out that there is no evidence that any of the GE food products currently approved for sale in the United States have been harmful to consumers. But it also said that the testing procedures used to determine the safety of these products are not ideal.

Despite all the worries, many people see GE food as a blessing. Many of the same things that environmentalists see as problems, others see as improvements. For instance, because farmers can insert a gene into a plant that makes it resistant to herbicides, they can kill weeds without killing their crop. "I think there is so much promise in this technology," says one expert at the Council for Agricultural Science and Technology. "There are some questions out there that we don't have the answers to, but once we see the health and nutritional benefits come out I don't think we'll see the opposition by the public."

Some of those benefits are right around the corner. Low-fat french fries from potatoes that have been engineered to require less oil for frying; rice with higher levels of lysine, which is essential for good health; and fruits and vegetables with increased amounts of nutrients like beta-carotene and vitamins E and C are all in the works. Products already on the market include corn that will grow in tough soils, soybeans that produce oil with higher levels of oleic acid and lower levels of saturated fats, peppers with improved taste and color, and firmer, tastier, more colorful tomatoes. Some believe these products and others are the world's answer to a growing population faced with decreasing amounts of farmland. Crops that can produce their own pesticides, resist common diseases, survive the spraying of popular herbicides, and provide more nutrition per acre can only help, they argue.

The debate over the safety of genetically engineered foods will probably continue for a long time, as people on both sides of the issue have valid points. As it is now, the only way you can tell for sure that the foods you buy at the supermarket are not genetically engineered is if you buy certified-organic food. By law, certified-organic food cannot contain genetically engineered ingredients. On the other hand, if the thought of GE foods in your diet doesn't bother you, eat whatever foods you like and don't worry about it.

If you have an opinion and want to be heard by the government, write a letter to your representative in

Congress (http://www.house.gov/writerep). Tell him or her what you think and why, and take pride in the fact that you stood up for your beliefs.

The world of additives is changing every day. New additives are invented to meet new consumer demands—things like fat substitutes, sugar substitutes, and protein boosters. Old additives are reexamined and reevaluated for safety. New technologies are created that allow companies to create more realistic flavors and more realistic colors. New studies emerge claiming this substance or that substance causes cancer. Other studies emerge saying the same substances are the next best way to stay healthy and live a long and happy life.

The controversies and arguments will never end. As long as there are big companies creating new foods, and as long as people have the right to ask what is in those foods, the debates will continue. There are many uncertainties, and some may never be resolved. Whatever happens, however, one thing is for sure: Whether you eat additives or not is really up to you.

Glossary

antibrowning agent A chemical used to preserve food and prevent food from becoming brown.

anticaking agent A substance added to a food to keep its parts from clumping together.

antimicrobial agent A substance added to food to prevent the growth of potentially harmful microorganisms.

antioxidant A substance that prevents or slows the oxidation of food ingredients, a process that eventually makes food go bad. Antioxidants protect against cancer-causing free radicals.

carcinogen Any substance that causes cancer.

deficiency A shortage of a particular substance or nutrient that may be necessary for good health.

dye A kind of color additive used primarily in moist foods like drinks, baked goods, and dairy products.

emulsifier A substance added to food to help spread flavors and oils throughout the entire product.

FDA The U.S. Food and Drug Administration; a government agency that ensures the safety of the nation's food supply.

flavor enhancer A substance added to food in order to change the way it tastes or make its natural flavor more noticeable.

food additive A substance added to food to improve flavor, appearance, or nutritional value.

Good Manufacturing Practices FDA regulations that set a limit on the amount of additives allowed in certain foods.

GRAS Generally recognized as safe; a category of additives that includes substances the FDA believes to be safe for human consumption based on scientific evidence or a long history of safe use.

lake A kind of pigment used to add color to foods.

leavening agent A substance added to dough to make it lighter in weight.

nutrients Substances found naturally in many foods that are essential for good health.

packaging material Materials like plastic, cardboard, and paper used to hold food before it is eaten.

palatability How good a food tastes.

pigment A coloring substance used in the food-manufacturing process.

preservative An additive used to protect a food against decay, discoloration, or spoilage.

processed foods Foods that are created through mixing, blending, and treating of whole foods, chemicals, and other substances.

stabilizer A substance added to processed food to keep it in a certain condition.

sweetener A substance added to food to make it sweeter than it naturally is.

synthetic Produced artificially.

texturizing agent A substance added to processed food to give it a certain texture.

toxic Harmful or poisonous.

vitamin-fortified Containing vitamins that have been added to increase a food's nutritional value.

Where to Go for Help

American Academy of Allergy
Asthma & Immunology
611 East Wells Street
Milwaukee, WI 53202
(800) 822-2762
Web site: http://www.aaaai.org

American Dietetic Association
216 West Jackson Boulevard
Chicago, IL 60606-6995
(312) 899-0040
Web site: http://www.eatright.org

Center for Science in the Public Interest
1875 Connecticut Avenue NW, Suite 300
Washington, DC 20009
(202) 332-9110
Web site: http://www.cspinet.org

Community Nutrition Institute
910 17th Street NW, Suite 800
Washington, DC 20006
(202) 776-0595
Web site: http://www.unidial.com/~cni/
homepage.htm

Consumer Federation of America, Food Policy Institute
1424 16th Street NW, Suite 604
Washington, DC 20036
(202) 387-6121
Web site: http://www.consumerfed.org/
backpage/fpi.htm

International Food Information Council
1100 Connecticut Avenue NW, Suite 430
Washington, DC 20036
(202) 296-6540
Web site: http://www.ific.org

National Food Processors Association
1350 Eye Street NW, Suite 300
Washington, DC 20005
(202) 639-5900
Web site: http://www.nfpa-food.org

National Institutes of Health
Visitor Information Center
Bethesda, MD 20892
(301) 496-4000
Web site: http://www.nih.gov

Union of Concerned Scientists
2 Brattle Square
Cambridge, MA 02238-9105
(617) 547-5552
Web site: http://www.ucsusa.org

United States Department of Agriculture
Center for Nutrition Policy and Promotion
1120 20th Street NW
Suite 200, North Lobby
Washington, DC 20036
(202) 418-2312
Web site: http://www.usda.gov/cnpp

U.S. Food and Drug Administration
Center for Food Safety and Applied Nutrition
200 C Street SW
Washington, DC 20204
(888) SAFEFOOD (723-3366)
Web site: http://vm.cfsan.fda.gov

World Health Organization
Regional Office for North America
525 23rd Street NW
Washington, DC 20037
(202) 974-3000
Web site: http://www.who.org

For Further Reading

Farlow, Christine H. *Food Additives: A Shopper's Guide to What's Safe & What's Not*. Rev. ed. Escondido, CA: KISS for Health Publishing, 1999.

Garrison, Robert H., and Elizabeth Somer. *The Nutrition Desk Reference*. 3rd ed. New Canaan, CT: Keats Publishing, 1995.

Renders, Eileen. *Food Additives, Nutrients, and Supplements A-to-Z: A Shopper's Guide*. Santa Fe, NM: Clear Light Publishers, 1998.

Sarjeant, Doris, and Karen Evans. *Hard to Swallow: The Truth About Food Additives*. Blaine, WA: Alive Books, 1998.

Schlosser, Eric. *Fast Food Nation: The Dark Side of the All-American Meal*. Boston: Houghton Mifflin Company, 2001.

Simontacchi, Carol. *The Crazy Makers: How the Food Industry Is Destroying Our Minds and Harming Our Children*. New York: J. P. Tarcher/Putnam, 2000.

Twogood, Daniel A. *MSG Is Everywhere*. Victorville, CA: Wilhelmina Books, 1996.

Winter, Ruth. *A Consumer's Dictionary of Food Additives*. 5th ed. New York: Three Rivers Press, 1999.

Wood, Rebecca. *The New Whole Foods Encyclopedia: A Comprehensive Resource for Healthy Living*. New York: Penguin/Arkana, 1999.

Index

About the Author

Chris Hayhurst is a freelance writer living in Colorado.

Photo Credits

Cover, pp. 2, 19, 49 © International Stock; pp. 7, 9, 13, 23, 26, 31 © Index Stock; p. 39 © Cindy Reiman.

Designer

Nelson Sá